PIPEWORK AND BRAZING

REFERENCE MANUAL

R6

Published by ConstructionSkills, Bircham Newton, King's Lynn, Norfolk, PE31 6RH

© **Construction Industry Training Board 1999**

The Construction Industry Training Board otherwise known as CITB-ConstructionSkills and ConstructionSkills is a registered charity (Charity Number: 264289)

First published 1999
Revised 2001
Revised April 2003
Revised June 2004
Revised July 2007

ISBN: 978-1-85751-249-6

ConstructionSkills has made every effort to ensure that the information contained within this publication is accurate. Its content should be used as guidance material and not as a replacement for current regulations or existing standards.

All rights reserved. No part of this publication may be reproduced, stored in a retrieval system or transmitted in any form or by any means, electronic, mechanical, photocopying, recording or otherwise, without the prior permission in writing from ConstructionSkills.

Printed in the UK

For a comprehensive listing of all BES publications turn to the back page.
Tel: 01485 577800   Fax: 01485 577758   E-mail: publications@cskills.org

CONTENTS

                                                       Page

ACKNOWLEDGEMENTS

FOREWORD   1

1. MATERIALS   2
   1.1 Brazing   2
   1.2 Brazing Alloys (Rods)   2
   1.3 Copper to Copper Joints   2
   1.4 Copper to Non-Ferrous Joints   3
   1.5 Silver Brazing Alloys (Cadmium Bearing) Grades   3
   1.6 Flux   3

2. HEALTH AND SAFETY   3
   2.1 Hazards   3
   2.2 Personal Safety   4
   2.3 Goggles, Face Shields and Safety Glasses   4
   2.4 Gloves and Arm Protectors   4
   2.5 Overalls   4
   2.6 Head and Foot Protection   4
   2.7 COSHH (Control of Substances Hazardous to Health) Risk Assessment and Procedures   4
   2.8 ACOP (Approved Code of Practice)   5
   2.9 Permit to Work   5
   2.10 Pressure Systems and Transportable Gas Containers Regulations   5
   2.11 Risks and Hazards   5
   2.12 Ventilation   6
   2.13 Cadmium   6
   2.14 Equipment Safety   7

3. BRAZING AND PIPEWORK   7
   3.1 Annealed Soft Copper Tube   8
   3.2 Half-Hard Copper Tube   8
   3.3 Cold Drawn Hard Copper Tube   8
   3.4 Pipe Joints   8
   3.5 Protection   8
   3.6 Types of Joint   8
   3.7 Joint Preparation   9
   3.8 Pipe Cutting   9
   3.9 Joint Clearance   9
   3.10 Swaging Soft Coil or Annealed Copper Tube   10
   3.11 Joint Cleanliness   10
   3.12 Joint Alignment   10
   3.13 Joint Fluxing   11
   3.14 Brazing   11
   3.15 Completion   12
   3.16 Finishing Off   12
   3.17 Thermal Insulation   13
   3.18 Flare Connections   14
   3.19 Making a Flared Connection   14

(continued overleaf)

	Page

4. EQUIPMENT PREPARATION   15
    4.1 Gas Cylinders   16
    4.2 Acetylene Cylinder   17
    4.3 Oxygen Cylinder   18
    4.4 Hoses   18
    4.5 Regulators   18
    4.6 Blowpipe   19
    4.7 Nozzles   19
    4.8 Purging Air from Pipework   19

5. LIGHTING PROCEDURE   20
    5.1 Acetylene   20
    5.2 Lighting the Torch   21
    5.3 Types of Flame   21
    5.4 Backfire   23
    5.5 Flashback   23
    5.6 Setting Regulators   23
    5.7 Shutting Down the Plant   24
    5.8 Fire Precautions   24

ACKNOWLEDGEMENTS

ConstructionSkills wishes to thank staff of the organisations below for their assistance in the production of this reference manual.

Consultants

- HAL Training Services
- Polar Pumps Training

Others

- Murex Welding Products Ltd
- Techtrain Associates
- Training by M.E.

FOREWORD

The need for reliable brazing procedures has always been important within the refrigeration and air-conditioning installation industry. This is a point well recognised and endorsed by manufacturers of equipment and components.

The environmental problems caused by fluorocarbon refrigerant leakages are very serious.

> 'At least 70% (about 8,000 tonnes) of the refrigerant sold annually in the UK is used to re-charge systems that have lost refrigerant through leakage'.*

The costs incurred by increased power consumption due to leakage must also be considerable.

Refrigeration and air-conditioning equipment using 'high pressure' refrigerants (such as R410a) are at particular risk of leakage if brazed joints are not of the highest quality and if materials of the correct specification are not used. The running pressures and, therefore, test pressures needed are significantly higher (i.e. At 55°C, R22 = 20.75 bar g, R410a = 33.3 bar g.)

This point was emphasised, in 1999, by a warning and press release issued by the Air Conditioning and Refrigeration Industry Board (ACRIB), and backed by Guidance Note 7 issued by the Institute of Refrigeration.

BS EN 378 (2000), the British Refrigeration Association Specifications and the forthcoming European Pressure Equipment Directive all emphasise the need for competence in brazing.

Assessments for this course in Pipework and Brazing are competence based. Operatives must demonstrate the ability to undertake the work to a recognised standard. The standard used is the British Refrigeration Association (BRA) Jointing of Copper Pipework for Refrigeration Systems. Test pieces will be destructively tested to ensure good workmanship.

The success of operatives in achieving the assessment standards also enables employers and installation contractors to demonstrate their corporate competence and compliance with regulations.

Poor pipework bending, routing, installation and lagging are obvious to all craft disciplines who take pride in their work. Poor workmanship stands as an epitaph to the bodger (if it lasts) for years.

Unfortunately, poor joint penetration cannot be seen and is inherently tragic. If all engineers aim for perfection in the standard of their brazing – what a difference it will make to refrigerant leakage and the costs incurred by industry.

Based on the British Refrigeration Association (BRA), Level 2 and 3 Specification of Procedures for Manual Flame Brazing and Brazer Assessment.

* Institute of Refrigeration's Code of Practice for the Minimisation of Refrigerant Emissions from Refrigerating Systems (ISBN 1 872719 06 6)

1. MATERIALS

1.1 Brazing

In recent times the term 'brazing' has been used to refer to several distinct forms of jointing. In fact, brazing is only one form of hard soldering and should really only be applied when using a filler rod of brass (copper and zinc) alloy.

We cannot, however, ignore the fact that many people now use the term 'brazing' for all hard soldered joints used in fluorocarbon pipework installation.

The British Refrigeration Association document *BRA Specification of Procedures for Manual Flame Brazing and Brazer Assessment* defines brazing as:

> 'A process of jointing generally applied to metals in which, during or after heating, molten filler metal is drawn by capillary action into and retained in the space between the closely adjacent surfaces of the components being joined. In general the melting point of the filler material is above 500°C, but always below the melting point of the parent metal.'

1.2 Brazing Alloys (Rods)

These are the filler metals which, when molten, are drawn into the gap between the parts to be joined. When they have cooled, they form a bond between the joint surfaces.

Three groups of rods are generally used in fluorocarbon refrigerant pipework installation:

- copper/phosphorus/silver alloys (cadmium free)
- copper/silver brazing alloys (cadmium free)
- copper/silver brazing alloys (cadmium bearing).

They are used for joining the following types of joints:

1.3 Copper to Copper Joints

The rods used for this type of joint should be copper/phosphorus/silver alloys.

These are classified as self-fluxing, but flux can be used if required (usually with parent material with a copper content of less than 90%). The amount of silver (Ag) content varies from less than 0.5% to 15%, dependant on the manufacturer and the type of rods which are used.

1.4 Copper to Ferrous and Non-Ferrous Joints

For dissimilar metal joints (most usually copper to brass, or copper to bronze), **copper/silver alloys (cadmium free)** should be used. These are generally known as silver solder alloys (rods), and they require the use of flux.

The silver content typically varies between 15% and 56%. The higher silver-containing alloys in this group approach the level of properties exhibited by the cadmium bearing grades, but lack the exceptional fluidity of cadmium alloys. Some alloys in this group also contain tin (often around 2%) to improve performance.

1.5 Silver Brazing Alloys (Cadmium Bearing) Grades

The cadmium bearing group of alloys is the lowest melting and most free flowing of all the silver bearing categories. Health hazards presented by the emission of cadmium oxide fumes during brazing now curtails their use. Correct ventilation, personal hygiene and respiratory protection equipment (RPE) is a COSHH requirement.

Note: Cadmium bearing brazing alloys must not be used in food, medical and beverage applications.

1.6 Flux

This is a substance that will dissolve oxides and form a film that will prevent the formation of more oxides within the joint during the brazing operation. It will also keep the surfaces to be bonded clean during brazing. Fluxes commonly used for refrigeration pipework joints are normally in powder form and contain borate salts (which are acidic). The powder should be mixed to a smooth paste in accordance with the manufacturer's instructions.

Pre-fluxed rods are available in both silver solder and brazing rods. It would still be necessary to add flux paste before using pre-fluxed rods to ensure a clean, well penetrated joint.

On completion of the joint, flux residue should be removed because it will cause corrosion of the joint once in contact with atmospheric moisture.

2. HEALTH AND SAFETY

2.1 Hazards

Brazing is an inherently hazardous operation. The use of a high temperature flame has obvious dangers. The materials used in brazing can have less obvious but none the less important hazards and risks.

It is essential that operatives are fully aware of all the hazards involved in brazing. They must also take every precaution to minimise the risk and danger to themselves and others including visitors and contractors on their site.

2.2 Personal Safety

Personal protective equipment (PPE) manufactured since 1995 should bear the CE mark. The task should be 'risk assessed' and the users consulted to ensure the suitability of PPE.

Operatives must protect their head, eyes, body and clothing from the effects of heat and hot metal.

2.3 Goggles, Face Shields and Safety Glasses

Eyes can be affected by heat, hot particles and glare. It is essential that appropriate eye protection be worn to provide protection against these risks. One pair of eyes is your allocation – keep them safe from harm.

2.4 Gloves and Arm Protectors

Heat-resisting gloves should be worn to protect hands from heat and hot particles. Some fluxes may cause skin irritation. Suitable gloves or barrier creams should be used when applying flux. When working in a restricted space, arm protectors may be required. Burns are extremely painful and cause scarring – make sure your skin is protected.

2.5 Overalls

Fire retardant or heavy cotton overalls with long sleeves should be worn when brazing and preparing joints. Synthetic materials are likely to burn. Nylon melts and sticks to the skin. Both should be avoided.

2.6 Head and Foot Protection

When working where there are risks from hot components or falling objects, head and foot protection may be required. Hard hat and steel toecaps – your best friends.

2.7 COSHH (Control of Substances Hazardous to Health) Risk Assessment and Procedures

All the materials used in brazing have risks attached and are hazardous. Before undertaking work, operatives must become familiar with data sheets and risk assessments for brazing gasses, filler materials and fluxes. Manufacturer's/supplier's data is not generally acceptable on its own. Companies should review this information and put it into context to suit their own use. If materials are particularly hazardous, then operatives must be trained to minimise risks.

2.8 ACOP (Approved Code of Practice)

It is now widely believed that any work practice that is not in line with an approved code of practice is considered as being negligent. An approved code of practice (which, in effect, is a set of guidelines) is issued by the Health and Safety Executive (HSE). Other organisations such as BRA, ACRIB and the Institute of Refrigeration also issue their own codes of practice.

Where applicable, company procedures should be familiarised and adhered to. An operative's own company procedure and those of customer companies must be followed.

2.9 Permit to Work

In order to minimise risks, companies should devise a safe system of work based upon their risk assessments. These must be documented if the company employs more than five people.

The most formal of these is a Permit to Work. This could be either the company's own or the customer's. When carrying out this type of work, companies/contractors should communicate risks to the customer/client. Customers/clients should also communicate any particular hazards to their contractors.

2.10 Pressure Systems and Transportable Gas Containers Regulations

The Pressure Systems and Transportable Gas Containers Regulations 1989 define how gas cylinders must be stored and transported.

2.11 Risks and Hazards

Operatives must take their own health and safety decisions but the following should be considered:

Local permission and isolation of alarms and detectors

Operational and production schedules may be affected. Intruder, fire, flame or movement detectors may need to be isolated.

Suitable fire extinguisher to hand

For example, carbon dioxide (CO_2) or dry powder. Make sure that fire extinguishers are in date and are serviceable for use.

Personal injury from burns, fumes, heat, glare, particles, flux and oxygen depletion

Use of heat shields, tinted eye protection or face shield, extractor or vent fan, barrier cream and gloves. The area should be well lit using a flame-proof hand lamp or torch where necessary.

Fires or explosion due to oxygen or acetylene leaks

Leak test all joints and seals between the cylinder and blowpipe.

Backfires and flashbacks of flame into blowpipe or hoses

Check cylinder contents regularly. Deal with poor flame pattern by shutting down to clean nozzle and reset flame.

Setting fire to adjacent material

Clear work area of flammable debris before starting work. Shield flammable adjacent material which may be the cause of inadvertent fires.

Exclude bystanders from viewing

Danger from all of the above unless properly protected, i.e. under strict training situations.

2.12 Ventilation

Brazing can often produce harmful fumes. These have to be carried away from the work area quickly. Adequate natural ventilation must be ensured if brazing is to be carried out.

In areas where natural ventilation is not adequate, or is not evident, then Local Extract Ventilation (LEV) must be employed. Either fresh-air fans or, preferably, exhaust fans should be used.

Care should be taken that by exhausting contaminated air, it does not cause problems in another location.

There should be NO BRAZING where leakage of refrigerant has occurred until the area has been ventilated and is safe. Wherever adequate ventilation cannot be ensured, breathing apparatus must be used.

As stated, fumes are given off by normal brazing operations. However, overheating of fluxes can be very hazardous. When fluxes are overheated they can give off toxic fumes.

Adequate lighting, natural or supplementary, should be available.

2.13 Cadmium

Some brazing rods contain cadmium. When cadmium is heated to above 320°C then toxic cadmium oxide fumes are given off. Not only can these cause acute, immediate respiratory problems but also long-term health may be affected.

It is essential for operatives to ensure that, wherever possible, cadmium-free alloys are used. If this is not possible then all necessary steps to control exposure must be taken.

Note: In food industry applications cadmium rods MUST NOT BE USED. This also applies to medical breathing gasses and beverage lines.

2.14 Equipment Safety

It is vital that all equipment is in good condition and used in a proper and safe manner (see *Safe Under Pressure*, published by BOC).

- Flashback arrestors and check valves must be in place.

- Nozzles, torches, hoses, regulators, flashback arrestors and cylinders must be properly inspected on a regular daily basis.

- Cylinders must be properly secured in an upright position – this includes inert gas cylinders.

- Gas pressures must be correctly set at the regulators. The correct flame cannot be safely achieved by manipulation of the torch valves only.

- Suitable and adequate fire-fighting equipment must be readily available throughout the brazing process. Suitable and adequate heat shields must be used to protect any surrounding material.

3. BRAZING AND PIPEWORK

PLUMBING GRADE AND SIZE COPPER PIPE AND COMPONENTS MUST NOT BE USED IN ANY CIRCUMSTANCES.

The reason for this is that the thickness of the wall of the piping is not suitable for refrigeration pressures.

It must be stressed that the internal cleanliness of refrigeration pipework is of paramount importance. It is this pipework which connects all the system components. Dirt and moisture are most likely to enter the system during the installation of pipework.

Note: Every precaution must be taken to ensure that dirt, air and moisture do not enter the system.

All installations which use fluorocarbons as refrigerants use copper piping. This copper piping must be seamless, deoxidised and dehydrated.

It is designed as 'refrigeration quality' to BS 2871-2: 1972, ASTM B280-86 and ASTM B88.

It is available in either cold drawn hard, half hard or annealed condition in sizes from $\frac{1}{8}$" outside diameter (OD) to $4\frac{1}{8}$" OD, and 12 to 22 standard wire gauge (swg).

Metric refrigeration quality copper tubing is available from your supplier.

3.1 Annealed Soft Copper Tube

Annealed soft copper tube is supplied in sizes from ⅛" to ⅞" OD, 19 to 22 swg, in 6, 10, 15 and 30 m coils with ends crimped or sealed to preserve dryness. Tubing should be cut from one end of the coil. The remainder is then resealed. This tube is pliable and easily bent by hand using external tube benders but can look unsightly unless care is taken during installation. The tube should be unrolled from the coil to the length required on a flat bench or level surface. The length required is cut with a wheel cutter and the coil re-sealed. No attempt should be made to re-roll the coil as this 'work hardens' the copper, causing unnecessary stress points. In unrolling the coil, it becomes slightly elliptical in cross-section so it may need the leading end of a swage punch to correct it after de-burring. This tubing can be swaged or flared. Flared fittings and components from ¼" to ¾" diameter are available. Soft copper tube can be "formed" using bending pliers or "external" bending springs to prevent "flattening".

3.2 Half-Hard Copper Tube

Half-hard copper tube is supplied in 3 m and 6 m straight lengths from ⅜" to 2⅛" OD, 14 to 20 swg, with ends sealed to preserve dryness. Tubing should be cut from one end of the tube with a wheel cutter and the remainder resealed.

3.3 Cold Drawn Hard Copper Tube

Cold drawn hard copper tube is supplied in 3 m and 6 m straight lengths from 2⅝" to 4⅛" OD, 14 to 18 swg, with ends sealed to preserve dryness. Tubing should be cut from one end of the length with a wheel cutter or, for the larger sizes, a skillfully-used, fine-toothed hack saw (32 tpi) or bandsaw and the remainder resealed.

3.4 Pipe Joints

A wide range of pipe fittings is available for use with half-hard and cold drawn hard refrigeration tubing. These fittings may be connected to the tubing by silver brazing. Precautions must be taken to prevent dirt and moisture from entering the fittings.

3.5 Protection

Straight lengths are supplied with the ends plugged. Coils are also plugged, then sealed in plastic bags to preserve dryness. Joints are supplied individually wrapped in plastic bags for the same reason.

3.6 Types of Joint

There are two methods of making joints to connect pipes and components used in air-conditioning and refrigeration systems:

- mechanical joints
- brazed or soldered joints.

While it will not be possible to totally eliminate the use of mechanical joints, these should be avoided where practically possible. This will minimise the risk of leaks from, what are after all, temporary connections, due to corrosion, expansion and contraction.

3.7 Joint Preparation

Hard and half-hard drawn piping should be used with connecting joints and allowed to cool in air to maintain its half hardness. Hard drawn has a slower cooling period because of its mass. (Do not assist cooling as this will anneal the copper.)

Soft roll copper tube is annealed and can be swaged for joints but the joining of soft/hard copper should be avoided where possible. If joints are allowed to cool in air, then that area becomes half-hard and stress areas are created each side of the joint. Annealing is achieved through quenching with water from bright red.

In all cases, careful attention to joint preparation is essential, enabling the formation of a sound joint first time. This will save time and prevent costly failures.

Brazing joints are either formed using a swaging/flaring tool or utilise manufactured fittings. Fittings and pipework maybe similar metals (i.e. copper/copper) or the parent material may be dissimilar (i.e. copper/brass or copper/steel).

3.8 Pipe Cutting

Pipes should be cut using a wheel type pipe-cutting tool. Ensure that the blade is sharp and not damaged. Ream and de-burr the cut.

Ensure that swarf does not enter the tubing by holding it downwards. Tap the piece to dislodge shavings.

Do not blow into the pipe. This is a natural action by an engineer but will introduce all the adverse effects to the interior of pipework caused by moisture.

A thin smear of compressor oil when cutting, reaming, de-burring and bending will be beneficial, but it must be cleaned thoroughly before brazing.

A swaging set, flame equipment, supporting jigs/frames, tongs (for handling hotwork), measuring tape/rule and an appropriate fire extinguisher should be available and to hand before work commences.

3.9 Joint Clearance

When you are using preformed manufacturer's fittings the joint gap or clearance will be between 0.05 mm (2 thou) and 0.25 mm (10 thou) depending on the diameter of the pipe. If the gap in the joint is outside these dimensions, capillary action will not take place. Where a manually formed swage is used using a punch, the clearance should not exceed 0.2 mm. A more accurate clearance can be achieved using adjustable swaging tongs. The joint clearance should be even. All pipe irregularities should be removed. The depth of the finished swage should be at least one pipe diameter.

3.10 Swaging Soft Coil or Annealed Copper Tube

The length to be swaged should be mounted in the swaging/flaring stock or bar and should protrude one and one-eighth (1⅛) times the diameter of the pipe. This allows for shrinkage when the pipe is being stretched. The dolly should be lightly smeared with compressor oil (to ease penetration). Then the inside of the pipe cleaned with degreaser after completion. Ensure that the dolly is square to the pipe and bar then proceed with the swage.

The drawing below shows two typical faults caused by poor tube protrusion.

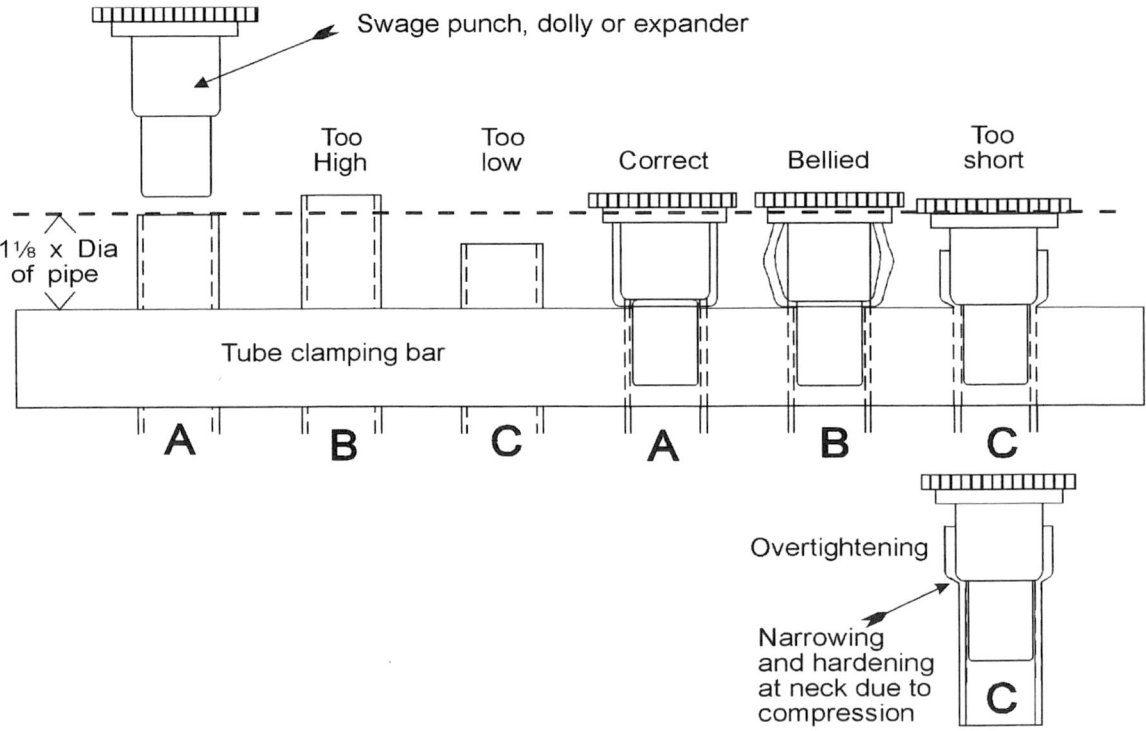

Faults caused by poor tube protrusion

3.11 Joint Cleanliness

The importance of removing all scale tarnishing, oxide, grease, oil or any other contamination cannot be over-emphasised. Time spent on ensuring joint surfaces within and adjacent to the joint are very clean will be more than paid back later in the brazing process.

Fumes from oil, paint, grease or other contaminants burning off near to the joint can prevent the proper filling of the joint.

3.12 Joint Alignment

Correct alignment should be ensured throughout the brazing. The wiring together of joints and the use of jigs should be employed as necessary.

3.13 Joint Fluxing

Where fluxes are necessary these should be applied to all surfaces during assembly. Male components must be engaged to female by 2 mm before fluxing. This will prevent flux intruding to leading edge of pipe joint. Powder fluxes are generally mixed to a paste.

Remember to protect your skin and eyes from any possible detrimental effect of the flux. Goggles, gloves and barrier creams should be used as necessary. Care must be taken to ensure complete coverage, but not to have excess flux in the pipe or joint. After pre-fluxing and assembly, excess flux should be removed.

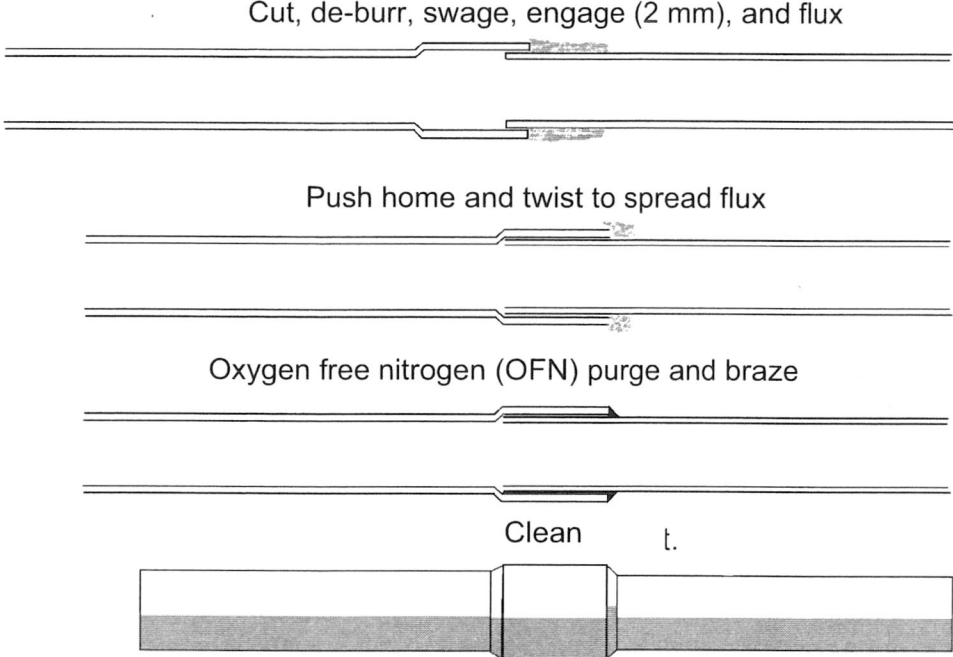

Fixing and brazing joints

3.14 Brazing

Heat the joint in an even manner by keeping the flame moving. Play the flame on the heavier or more conductive parts more often. Try to keep the joint in the hottest part of the flame but away from the tip of the inner cone (or you'll have a 'one note' penny whistle).

Remember!

- The filler rod should be melted by the heat in the joint, not by the flame.
- The filler rod should be drawn into the joint by capillary action, not pushed into the joint.

Do not bring the filler rod to the joint until the joint is near to the melting temperature of the rod. This can be determined by the colour of the parent material and sometimes by the liquefaction of the flux.

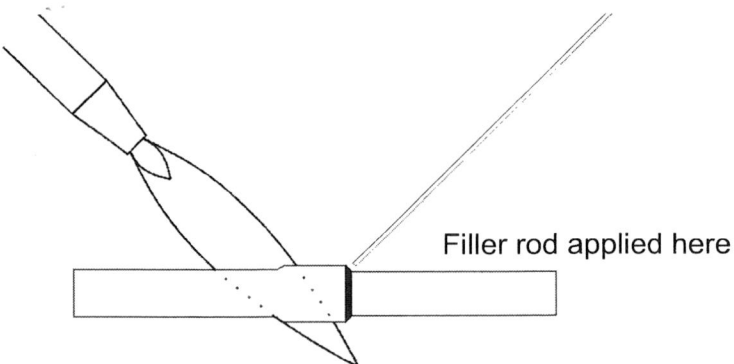

Position of flame and filler rod at joint

When the rod starts to melt as it is gently touched onto the joint, keep the flame moving but play the flame more to the base of the joint. This will promote the flow of alloy deep into the joint.

Similarly, play the flame around the joint to promote alloy flow all round the joint inner surface. Keep applying heat and filler rod until the joint is completely full and a bead of filler alloy is visible around the outer edge of the joint.

Take care not to overheat the joint. Overheating will oxidise the copper and possibly collapse the pipe. Draw the flame back from the joint if necessary. Insufficient heat will cause the joint not to be filled (capillary action will not take place).

3.15 Completion

After the joint has cooled remove all flux residues and visually inspect the joint. A proprietary plastic, or any non-metal cleaning pad may be used to clean the joint externally. Sandpaper, emery cloth and similar abrasives should not be used as they may reduce the tube wall thickness. The use of wire wool to clean the joint leaves metal particulates on the soft copper, filler and surrounding area. This causes oxides to form at the site of the joint due to atmospheric moisture. Any electrically conductive metal swarf near electric/electronic components must be avoided.

3.16 Finishing Off

As a final protection, the joint should now be sprayed with paint or lacquer. A sheet of card or paper will shield surrounding components. This provides an effective barrier and protects the joint from atmospheric moisture and prevents corrosion.

3.17 Thermal Insulation

Manufacturers of thermal insulation materials provide varieties to suit most applications. The engineer will normally be directed by the installation drawings as to the specifications of the insulation to be used. The various types of insulation tube, adhesive and fixing material include: CFC free, fire retardant, wall thickness, etc.

The purpose of insulating refrigeration pipework is to remove atmospheric influences during operation such as condensate or ice forming on the suction line with subsequent puddles, damp patches and mould.

Before the advent of preformed insulation tubing, the suction and liquid lines were thermally whipped together using copper wire. There were six turns every foot and the twist was tucked between the pipes. This procedure acted as a heat exchanger, providing liquid at the metering device nearer to evaporation bubble point and less chance of liquid slugs in the suction line to the compressor. The pipes were then 'corked', lovingly bound with linen/hessian or muslin soaked in plaster of Paris. When dry, they were painted to protect them from the elements or to suit the decor.

Today the industry tends to lag the suction line only. In the case of heat pumps and VRV (Variable Refrigerant Volume) systems, both are lagged. The objective is still the same; to protect the low temperature line from external condensate or ice and subsequent consequences.

When lagging is left open to atmosphere at any point, and the pipe cools down, air (and inevitably moisture) is drawn into the lagging. The moisture then condenses on the pipe. When the pipe warms up, the dry cold air expands and vents to atmosphere. This 'breathing' leaves moisture and oxygen inside the lagging where, hidden from view, they form a layer of ice and copper oxide.

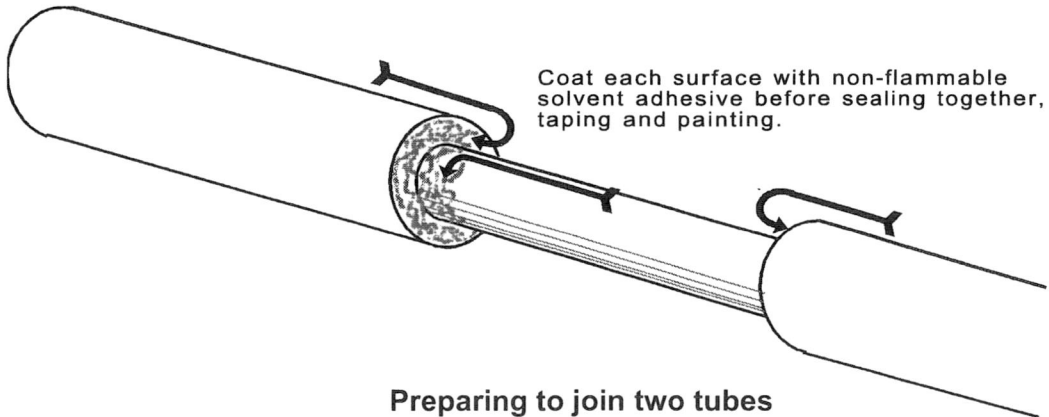

Preparing to join two tubes

Where the tube is to be joined to another tube, both must be sealed to the pipe and the butt ends of the tubes stuck together with non-flammable adhesive. Then they are taped to protect the joint. The open end should be stuck to the pipe to prevent breathing.

To protect the lagging from sunlight, birds and vermin, it should be painted with a couple of coats of water-based emulsion paint, to suit the interior/exterior decor, of course.

Other exposed components such as valves can be lagged with skilful use of hypafoam from an aerosol. This will prevent condensate drips. It can be chipped off to gain access and reveal a rust-free component at a later date.

3.18 Flare Connections

Flare connections can be used with soft copper tubing up to 3/4" OD. The fittings are normally made of brass and have a machined surface which mates with the flared end of copper tubing. The machined surface of the fitting must be free from any imperfections. A correctly made flare joint can be disconnected and re-joined during maintenance, but this practice is usually avoided due to vibration and temperature fluctuations making the joint unreliable.

3.19 Making a Flared Connection

- Measure the pipe length accurately to save material and give a neat job.

- Cut the tube to the required length with a wheel-type cutter to give the square end necessary for a flared connection.

- De-burr the tube with a knife or de-burring tool. Take care to prevent small pieces of copper entering the tube.

- To flare the tube, use a flaring tool consisting of a clamp and a cone. Place the tube in the clamp. It is important that the tube protrudes the correct amount through the clamp: too much and the pipe may split; too little and the flare will be too small to ensure a leak-proof connection. Smear the cone of the tool with refrigerant oil and screw it into the end of the tube until resistance is felt. Remove cone.

- To maintain the dryness and cleanliness of the tube, each piece of piping should be connected into the system as soon as it has been made.

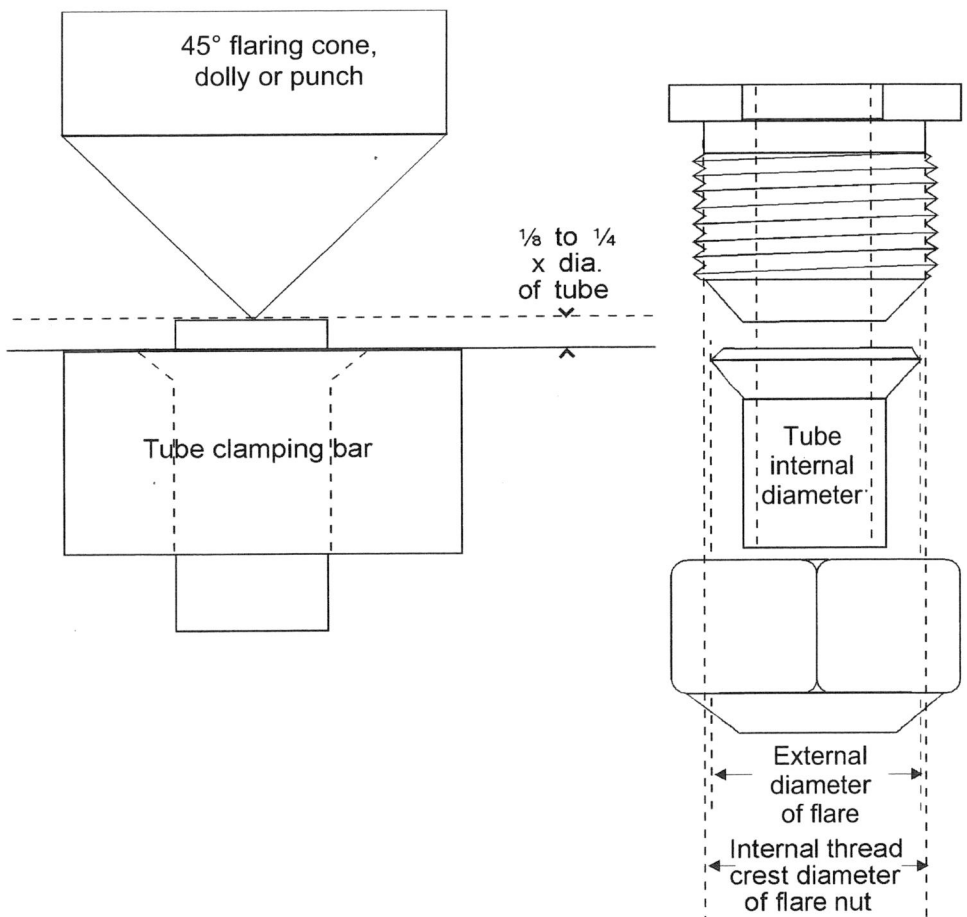

Making a flared connection

4. EQUIPMENT PREPARATION

All equipment should be thoroughly inspected prior to its use on a daily basis. All of the following equipment must be used.

4.1 Gas Cylinders

ACETYLENE – MAROON
OXYGEN – BLACK
OXYGEN FREE NITROGEN (OFN) – GREY, BLACK TOP, WHITE SPOT

- **Cylinder trolley** or method of securing all cylinders, including OFN.
- **Gas regulators** for all gases, clearly identified by name and colour.
- **Flow gauge** for purging gas.
- **Inert gas hose** (suitable material and colour code).
- **Regulator spanner** to secure gauges to cylinders in correct orientation.
- **Gas cylinder key** to open and close cylinder valve in one action (where appropriate).
- **Flashback arrestors** for oxygen and acetylene hoses.
- **Proprietary hoses,** no longer than necessary and in good condition.
- **Hose check valves** at torch end of each hose.
- **Torch** with correctly operating gas shank valves.
- **Nozzles** of appropriate size to work piece (change nozzle and reset regulators to increase/decrease heat).
- **Nozzle cleaners,** remove nozzles for cleaning.
- **Safety goggles** with filter lenses to protect from glare and fumes.
- **Gloves** to protect hands from heat and fluxes.
- **Ignitor, spark flint only.** If dropped, no secondary fires. Less danger of fingers contacting flame.
- **Suitable fire extinguisher,** read instructions before you start.
- **Heat shield** to protect other components, etc.
- **Joint preparation tools, equipment and materials.**
- **Suitable filler metal rods.**
- **Suitable flux.**

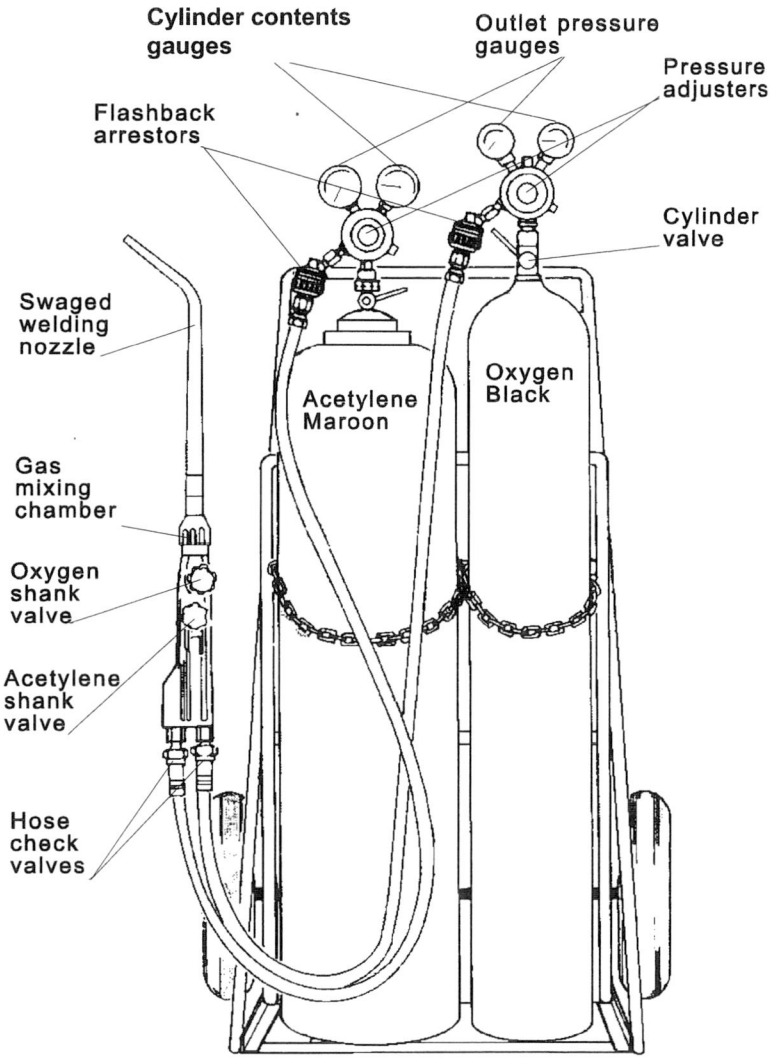

Acetylene and oxygen cylinders

4.2 Acetylene Cylinder

- An acetylene cylinder is filled with an absorbent, porous mass, e.g. kapok/lime silica.

- Acetylene gas is dissolved in acetone liquid.

- Acetylene, as a pure gas, is unstable and explosive above 2 bar g.

- Treat acetylene cylinders with great care, and always store and use in the upright position.

- Remember that acetylene equipment has a left-handed thread, this is indicated by grooves or chamfers marked on the connection nuts.

- If cylinders have been put on their side for any reason, even momentarily, they should be left upright for at least one hour to prevent acetone entering the regulator.

Note: Acetylene must not come into contact with components which have more than 70% copper content because it would produce explosive copper acetylides.

Acetylene must not be drawn out of the cylinder at a rate greater than one-fifth of the cylinder capacity per hour. This could cause acetone 'carry over' into the regulator, hose and torch.

4.3 Oxygen Cylinder

An oxygen cylinder is painted black, has right-handed threads and has a supply pressure of 230 bar g.

Note: Lubricants of any kind must not be used. Explosion may result.

4.4 Hoses

Particular attention should be given to hoses. They are the weakest link and most susceptible to leaking.

Make sure the connections are tight and that the hose is not abraded, cracked, cut, kinked or crushed. It is vital that the hose is suitable for the gas being transported.

You must only use proprietary oxy-acetylene hoses (or oxy-propane, etc.). They must be colour-coded:

BLUE – OXYGEN
RED – ACETYLENE

4.5 Regulators

Dual diaphragm regulators are preferable. The HP diaphragm drops the cylinder contents pressure to 7 bar. The LP diaphragm (controlled by the adjuster) controls the nozzle pressure more sensitively from 7 bar to 0.2 bar for most welding and brazing activities.

Ensure that regulators are in excellent order. Check that the pressure gauges for both cylinder contents and outlet pressure are sound and working correctly (they read zero when shut down). Regulators are colour coded with the gas for which they are to be used, clearly labelled.

Open cylinder valves slowly and carefully to prevent damage to regulators. The valve should be opened half to one turn so that it can be closed in one action.

4.6 Blowpipe

The blowpipe is also referred to as the torch or welding gun. This is normally a combination of shank, mixer and nozzle. Check that all connections are tight and sound. Also check that the valves are seating properly and operate smoothly.

4.7 Nozzles

Nozzles are known as tips. Make sure the nozzle orifices are clear and unobstructed. Always remove the nozzle from the torch for cleaning, otherwise the debris will fall back into the mixing chamber.

Nozzles are sized by number. No 1 is the smallest. No 5 or No 7 are ideal for pipe sizes between ¼" and ½" diameter. For higher or lower heat, change the nozzle.

Multi-hole nozzles (or pepper pots) have to be used on pipes between 1" and 4⅛" to reach the required penetration of the filler metal. It is not unusual to use two operatives with torches for large diameter pipes and components.

4.8 Purging Air from Pipework

It is essential that the pipework to be brazed is constantly purged of air before heating and during the brazing process with an inert GAS. Usually Oxygen Free Nitrogen (OFN) is used.

By purging the air (which contains oxygen) from the pipework the formation of copper oxides is eliminated.

If the air is not purged from the pipework, copper oxides are created on the inside while it is being heated. These are usually in the form of black scale. The scale can cause problems of blockages within the system, compressor oil contamination and subsequent wearing of bearing surfaces.

If refrigerants are not purged from the pipework and components, when heated they will produce extremely toxic, harmful acidic products.

Once all brazing is complete, it is important that purging continues until the pipework has cooled sufficiently to prevent the formation of scale.

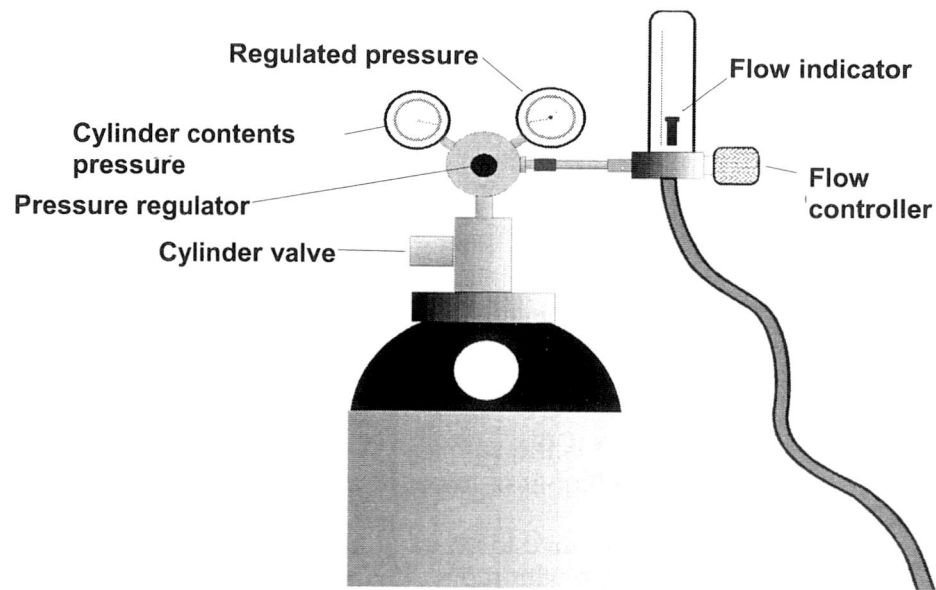

Oxygen Free Nitrogen (OFN) cylinder

An OFN cylinder fitted with a pressure regulator and flow gauge is connected to the pipework through a suitable hose.

The regulator is set to less than 1 bar g and the flow gauge to 2 LPM (litres/min) or 4.2 CFH (ft^3/hr) to allow OFN into the pipework. End caps with a 6 mm hole will ensure the same flow of OFN through each leg of the installation.

Ensure that the pipework does not become pressurised or the flow too great to the point where the brazing process is interfered with. Check the flow gauge regularly.

Too much pressure will blow the molten filler metal out of the joint. Too much flow will cool the joint internally.

Note: It is important that OFN is used and NOT industrial nitrogen. It could be argued that other purge gasses such as Argon could be used. These are, however, much more expensive than OFN. Whichever gas is used, ensure that NO oxygen is present in the gas.

BOC displays a **white spot** on the **black** collar of the **grey** OFN cylinder.

5. LIGHTING PROCEDURE

5.1 Acetylene

Due to its chemical structure, acetylene is the best fuel gas to use. A rapid break down in oxygen releases a large amount of heat in a short area of flame.

The hottest part of an oxy-acetylene flame is just in front of the inner cone and is approximately 3200°C. This temperature is much lower for propane and other fuel gases.

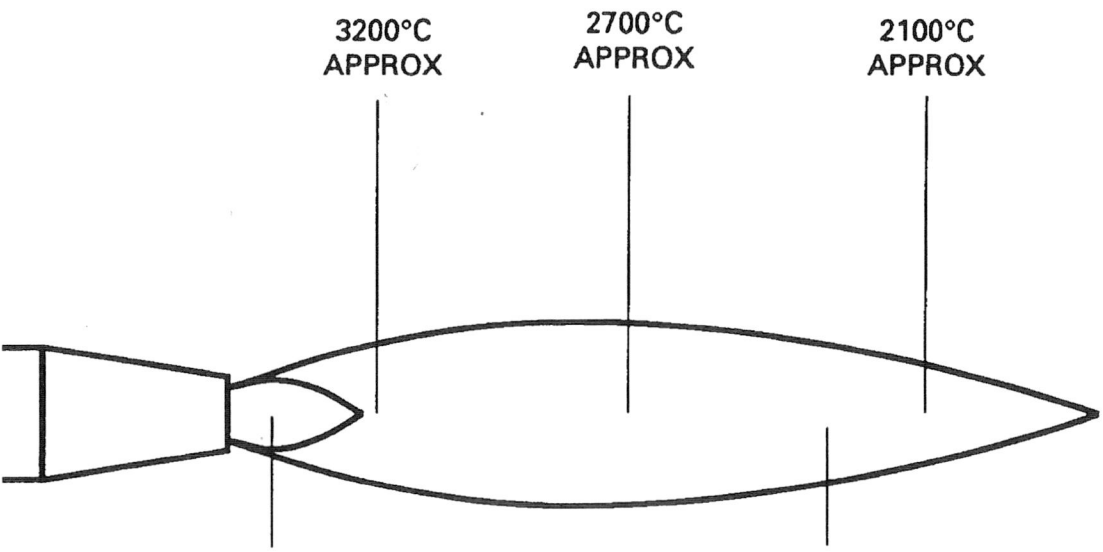

INNER CONE
Unburned gas inside the inner zone. Acetylene (C_2H) and oxygen (O_2), in equal volume, burn to produce the outer envelope

OUTER ENVELOPE
Carbon monoxide (CO) and hydrogen (H_2), which combine with oxygen from the surrounding air to produce water vapour (H_2O) and carbon dioxide (CO_2)

5.2 Lighting the Torch

Set up the regulators in accordance with 5.7. Open the acetylene valve on the blow pipe half a turn, then ignite the acetylene at the nozzle tip with a spark ignitor. A spark ignitor is recommended because it cannot cause secondary fires when discarded. When lit, continue to open the acetylene shank valve until the flame stops smoking. If a smoke-free flame cannot be achieved when the acetylene shank valve is fully open then increase the acetylene regulator outlet pressure slightly until the flame is not producing smoke. Now open the oxygen shank valve slowly and smoothly until a bright inner cone is clearly definable in the flame. This is 'neutral'.

5.3 Types of Flame

Three types of flame can be achieved with oxy-acetylene equipment: oxidising, neutral and carburising. A difference between the flame types can be seen by the overall size of the flame and shape and size of the inner cone.

- An oxidising flame has an inner cone that is distinctly pointed, the overall flame size is small and much bluer than other flames.

- A neutral flame has an inner cone with a rounded apex.

- A slightly carburising flame has a similar shaped inner cone with a visible slightly yellow flaring off the edges. A fully carburising flame has a defined second inner cone of yellowish tinge. Carburising flames have a larger overall flame size.

For brazing operations a neutral flame is best, but if the flame is to vary during use, a carburising flame is preferable to an oxidising flame which will 'burn' the components.

A carburising or oxydising flame is achieved by increasing or decreasing the oxygen supply to the smoke free acetylene flame.

For welding brasses and bronzes
Oxidising
Sharp inner cone, deeper colour in centre

For welding steel, cast iron, aluminium and copper
Neutral
Fully luminous inner cone

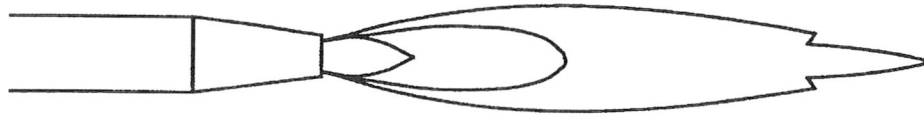

For hardfacing
Carburising

5.4 Backfire

A backfire is when the flame shoots back into the nozzle or blowpipe gas mixing chamber. It is not an uncommon occurrence and is not usually serious. Backfires can be caused by low gas pressure, dirty nozzles or by getting too close to the work. If a backfire occurs:

- turn off both blow pipe valves (oxygen first)

- release pressure from the regulators, acetylene first, then investigate the cause

- dismantle and clean the gas mixing chamber

- examine 'O' ring seals

- clean or replace the nozzle, then 'set up' the flame again (see 5.2 Lighting the Torch).

5.5 Flashback

A flashback is a serious event. A flashback is when the flame travels beyond the blowpipe check valve, into the hose or possibly the flashback arrestor. In this case the flashback arrestor will trigger and shut down the low pressure supply. If this happens:

- turn off the gas at the cylinders (oxygen first)

- thoroughly inspect all equipment to determine the cause and any damage resulting from the flashback

- do not reset the flashback arrestor/s until the cause of the flashback has been established and rectified

- turn of gases at cylinder valve.

5.6 Setting Regulators

After checking that the torch shank valves are closed, check that each regulator is closed by turning the adjuster fully anticlockwise. The cylinder valve may now be opened gently to prevent stressing the gauge (bourdon tube). Check that the regulator cylinder contents pressure gauge registers a usable contents pressure. Check/test for high pressure leaks using a proprietary leak detection fluid. DO NOT use soapy water as this may contain oil/soapy fats; grease and oxygen will explode.

After opening the appropriate blowpipe shank valve, turn the regulator outlet pressure adjuster to give a low delivery pressure reading. The pressure required will be quite low, in the order of 0.2 bar (3 psig).

Start by setting both fuel and oxygen regulator delivery pressures the same. Close the blowpipe shank valves after the air has been purged from each hose and check for low pressure leaks using a proprietary leak detection fluid. DO NOT use soapy water, this may contain oil/soapy fats; grease and oxygen will explode.

5.7 Shutting Down the Plant

Temporary shut down

- Close acetylene blowpipe shank valve (flame will extinguish).
- Close oxygen blowpipe shank valve.
- Stow hose and torch safely. DO NOT move the plant in this state*.

*If the plant topples during transit, it is very likely that the regulator/s will break off at the cylinder neck. This would result in the cylinder/s contents venting dangerously and uncontrollably.

Shutting down the plant

- Carry out 'temporary shut down' procedure.
- Close both cylinder valves.
- Open acetylene blowpipe shank valve until both acetylene pressure gauges read '0' psig and re-close.
- Open oxygen blowpipe shank valve until both oxygen pressure gauges read '0' psig and re-close.
- Back off both regulator hand wheels. Stow hoses and torch safely.

5.8 Fire Precautions

- Your company will already have in place procedures dealing with the hazard of fire. These will be in accordance with current health and safety regulations.
- They will also have carried out 'risk assessments' covering fire precautions.
- It is the duty of each engineer to carry out all precautionary measures stated by their company. This will ensure your own safety and the safety of your colleagues.
- The importance of following the rules and regulations prescribed by your company cannot be over-emphasised. Your own welfare and safety of others are of paramount importance.

BES PUBLICATIONS

Building Engineering Services continue to provide the gas, electric, water and refrigerant industries with a range of popular, respected and competitively priced publications.

These publications can be used either as the basis of training or for reference in the workplace. Some can also be used for assessment purposes. All are published in A4 format, with the most popular also available as A5, pocket-sized books.

DOMESTIC GAS

GAS SAFETY (G1) — Format: A4 in a ringbinder

The complete manual for reference or self-study. All of the essentials in 300 pages, with clear explanations and illustrations, covering ◆gas pipework ◆gas supply ◆combustion ◆appliance gas safety devices and gas controls ◆principles of gas flues ◆flueing standards ◆ventilation requirements ◆emergency procedures ◆unsafe situations ◆warning notices and labels. Also included is the HSE publication ◆*Safety in the installation and use of gas systems and appliances* (G31) which covers the HSE Gas Safety (Installation and Use) Regulations 1998 – Approved Code of Practice and Guidance, a ◆*Course Workbook* and a booklet of ◆*Practical Tasks* for you to complete.

GAS SAFETY (G2) — Format: A5 Wiro-bound

All the information and diagrams from *GAS SAFETY (G1)* in a handy size for reference on the job and for carrying in the service van.

DOMESTIC GAS APPLIANCES (G5) — Format: A4 in a ringbinder

Contains all seven of the domestic natural gas appliance manuals from ConstructionSkills in one package, plus the *Domestic Natural Gas Appliances Course Workbook (G14)*. The easy-to-use format makes it ideal for engineers working with a range of domestic appliances.

Each manual can also be purchased individually:
- Heating Boilers/Water Heaters (G7)
- Cookers (G8)
- Ducted Air Heaters (G9)
- Fires and Wall Heaters (G10)
- Tumble Dryers (G11)
- Meters (G12)
- Instantaneous Water Heaters (G13)

DOMESTIC GAS APPLIANCES (G6) — Format: A5 Wiro-bound

All the information and diagrams from the *DOMESTIC GAS APPLIANCES (G5)* in a handy size for reference on the job and for carrying in the service van.

FAULT-FINDING TECHNIQUES (G17) — Format: A4

Problems with locating that elusive fault? Follow the step-by-step techniques in this hands-on manual and speed up your fault finding on central heating systems.

SAFETY IN THE INSTALLATION AND USE OF GAS SYSTEMS AND APPLIANCES (G31) — Format: A4

An essential HSE publication for all those working with domestic gas. It gives advice on how to comply with *The Gas Safety (Installation and Use) Regulations 1998 – Approved Code of Practice and Guidance,* which has a special legal status. For example, if you are prosecuted for breach of health and safety law, and it is proved that you have not followed the relevant parts of the Code, a court will find you at fault (unless you can show that you have complied with the law in some other way).

COMMERCIAL AND INDUSTRIAL GAS

COMMERCIAL GAS SAFETY (G88) — Format: A4 in a ringbinder

An essential training and reference manual for those working in the commercial environment. It includes key sections from the popular GAS SAFETY (G1) and incorporates information from two other commercial publications (G23 and G24) which can be purchased separately) making this the definitive training and reference manual for commercial work. It covers ◆commercial gas safety ◆pipework and ancillary equipment ◆gas pipework ◆gas supply ◆combustion ◆appliance gas safety devices and gas controls ◆principles of gas flues ◆flueing standards ◆ventilation requirements ◆emergency procedures ◆unsafe situations ◆warning notices and labels. Also included is the HSE publication ◆*Safety in the installation and use of gas systems and appliances* (G31) and ◆*Course Workbooks* and *Practical Tasks* (G3, G4, G83 and G84).

COMMERCIAL GAS SAFETY (G23) — Format: A4

An essential supplement for engineers working in the commercial environment. If you already own a *GAS SAFETY (G1)* pack, all you need is this book with its commercial gas-specific sections ◆combustion and flue gas analysis ◆burners ◆controls and control systems ◆flues ◆ventilation ◆pressure and flow.

COMMERCIAL PIPEWORK AND ANCILLARY EQUIPMENT (G24) — Format: A4

An essential guide for engineers working on commercial pipework, with clear information on ◆pipework design ◆soundness testing and purging ◆commercial metering ◆boosters and compressors.

COMMERCIAL APPLIANCES (G25) — Format: A4

A comprehensive guide to the installation and commissioning of direct and indirect fired appliances, radiant heating and gas equipment.

COMMERCIAL CATERING (G26) — Format: A4

Essential information on installing, commissioning and servicing commercial catering appliances.

To obtain further information and order any of the publications listed, contact Publications on: Tel: 01485 577800 / Fax: 01485 577758 / E-mail: publications@cskills.org / www.cskills.org/publications

LIQUEFIED PETROLEUM GAS (LPG)

LIQUEFIED PETROLEUM GAS SAFETY (G80)
Format: A4 in a ringbinder/A4

The industry reference manual for those working only on LPG systems. It covers all you need to know about ◆combustion ◆appliance gas safety devices and gas controls ◆principles of gas flues ◆flueing standards ◆ventilation requirements ◆emergency procedures ◆unsafe situations ◆warning notices and labels.

This pack consists of: ◆*Gas Safety (G1) pack*, ◆*Liquefied Petroleum Gas Safety (G18) book*, ◆*Liquefied Petroleum Gas Safety Course Workbook (G81)*.

LIQUEFIED PETROLEUM GAS SAFETY (G18)
Format: A4

The essential bolt-on to those working with natural gas and looking to extend into LPG. If you already own a *GAS SAFETY (G1)* pack, all you need is this book with its LPG-specific sections ◆installation ◆fire precautions and procedures ◆combustion ◆testing and commissioning installations ◆service pipework ◆bulk gas supply systems ◆the leisure industry.

ELECTRICAL

BS 7671: REQUIREMENTS FOR ELECTRICAL INSTALLATION (E1)
Format: A4 Wiro-bound

The standard reference book for electrical work. The easy-to-follow text, supported by diagrams, explains the complex regulations in terms a practical electrician can understand. It now incorporates reference to the IEE on-site guide that enables you to make calculations and design circuits in a much quicker and simpler manner.

ELECTRICAL INSTALLATION PACK (E3)
Format: A4 in a ringbinder

Over 430 pages of illustrated reference material divided into four sections:
- Basic Practical Skills – describes the tools required for electrical installation work and how to use them
- Wiring Installation Practice – deals with terminating cables, flexible cords and installing PVC cables, conduit trunking, MICC, SWA and FP200 wiring systems. (Complies with the 16th Edition *IEE Wiring Regulations*)
- Basic Electrical Circuits – covers standard circuit arrangements for lighting and power circuits, and relevant IEE Regulations
- Safety at Work – essential advice on safety at work, from securing ladders to dealing with electric shock. It also gives the key points of relevant Acts and Regulations.

ESSENTIAL ELECTRICS (E14)
Format: A4

An indispensable reference book for plumbers, gas fitters and heating and ventilating engineers whose work requires basic electrical knowledge and an understanding of electrical regulations.

CENTRAL HEATING CONTROLS (E15)
Format: A4

Deals with different types of central heating control systems for wiring and fault finding.

COMBINATION BOILERS (E19)
Format: A4

An invaluable reference manual for engineers who want to understand the principles of combination boilers. This manual covers most of the content for the ConstructionSkills Essential Electrics and Combination Boiler Fault Finding course. Over 80 pages of illustrated reference information covering ◆types of boilers ◆designs ◆wiring diagrams ◆installation ◆commissioning and servicing ◆fault finding.

WATER

UNVENTED HOT WATER STORAGE SYSTEMS (W2)
Format: A4

An informative guide for installing unvented hot water storage systems. It covers most of the content for the ConstructionSkills training and assessment scheme, including: ◆types of system ◆design ◆controls ◆installation ◆commissioning and decommissioning ◆servicing and fault diagnosis ◆relevant Building Regulations ◆good practice.

REFRIGERANTS

SAFE HANDLING OF REFRIGERANTS (R2)
Format: A4

Essential information, primarily designed for operatives undertaking ConstructionSkills Safe Handling of Refrigerants training and assessments, it covers ◆environmental impact ◆fluorocarbon control and alternatives ◆regulations ◆recovery and handling ◆refrigeration theory ◆good practice ◆automotive installations.

SAFE HANDLING OF ANHYDROUS AMMONIA (R4)
Format: A4

Essential information for handling anhydrous ammonia. Primarily designed for operatives undertaking ConstructionSkills Safe Handling of Anhydrous Ammonia training and assessments, it covers ◆safety and environmental issues ◆regulations ◆good practice.

PIPEWORK AND BRAZING (R6)
Format: A4

Primarily for operatives undertaking ConstructionSkills Pipework and Brazing training and assessments for refrigeration systems, it covers ◆health and safety ◆materials and equipment ◆lighting procedures.

To obtain further information and order any of the publications listed, contact Publications on: Tel: 01485 577800 / Fax: 01485 577758 / E-mail: publications@cskills.org / www.cskills.org/publications